Fallow Deer
by Jochen Langbein and Norma Chapman

CONTENTS

Fallow Deer
By Jochen Langbein and Norma Chapman

Published by The Mammal Society and The British Deer Society

The Mammal Society	The British Deer Society
Registered Charity No. 278918	Registered Charity No. 228659
Registered Office:	*Registered Office:*
The Mammal Society	The British Deer Society
2B Inworth Street	Burgate Manor
London SW11 3EP	Fordingbridge
	Hampshire SP6 1EF

© The Mammal Society & The British Deer Society 2003

ISBN 0 906282 40 3
ISBN 0 902754 26 2

This is one of a series of booklets on deer in Britain published jointly by
The Mammal Society and The British Deer Society.

Drawings by Sarah Wroot.
Front Cover photograph and photograph on page 9 by Jon Jayes
Photographs on pages 15 by Ron Perkins; and pages 16 and 32 (bottom) by Russell Edwards
All other photographs by Jochen Langbein

Typesetting and printing by SP Press, Units 1 & 2, Mendip Vale Trading Estate, Cheddar Business
Park, Cheddar, Somerset.

Introduction

Fallow deer were present in Britain during the last interglacial period, but became extinct here and throughout most of Europe after the last glacial advance. They were re-introduced to Britain by the Normans during the 11th century and established in parks or royal hunting preserves. Numerous deer parks still flourish today but many lost their deer during the war years of 1914-1918 and 1939-1945 and were not restocked when peace was restored. A century ago only a few counties had any wild populations of fallow deer, but today they are one of the most widespread species of deer in the British Isles. Recent estimates suggest there may now be as many as 100,000 fallow deer living in the wild in Britain and a further 20,000 enclosed within parks or farms.

The attractive appearance of fallow deer, the excellence of the venison and the sporting interest have also been reasons why British and other European emigrants have transported this species around the globe. Wild populations are now established in such far-flung countries as Australia, New Zealand, Argentina, Antigua and Barbuda, South Africa, Fiji and Canada.

This booklet is about the European fallow deer, *Dama dama dama*. The only other living fallow deer is the Mesopotamian or Persian fallow deer, *D. d. mesopotamica*, a larger animal with antlers of a different form. In contrast to the European fallow, this sub-species is endangered with very few surviving in Iran. However, captive breeding programmes are running in several zoos, within Israel (where re-introductions to the wild have begun) and in New Zealand where techniques of artificial insemination and embryo transfer are being employed to increase the stock, using hybrid European x Mesopotamian females as surrogate mothers for Mesopotamian embryos.

Distribution

Some fallow are present in most counties of England and Wales and regions of Scotland, including the islands of Anglesey, Islay, Scarba and Mull. They are also found in four out of the six counties in Northern Ireland, as well as in Eire including Lambay Island. Although so widespread, particularly in England, and continuing to expand slowly, their distribution remains surprisingly localised, with many feral populations still established near to the many former and present day deer parks from which they escaped.

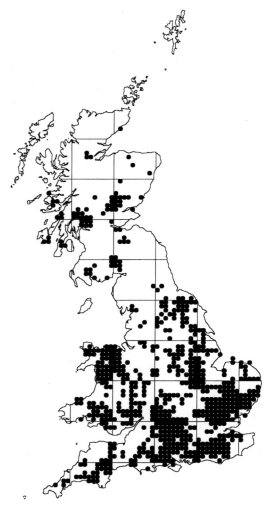

Distribution of fallow deer in England, Scotland and Wales (2001), Source: British Deer Society.

Appearance

Fallow deer occur in a variety of coat colours, which reflects many generations of breeding within parks where there was no natural selection against a particular type. The four main colours of fallow are known as common, menil, black and white, but there is seasonal variation too. Whilst the very spotted summer coat of the menil colour becomes less brightly spotted in winter, that of the common colour loses or almost loses its spots. The glossy summer coat of the black variety becomes dusky brown in winter. White fallow, so conspicuous in groups of mixed colours, are partial albinos: they have normal

Group of does and prickets ruminating; note the very wide range of coat colour variation among individuals.

eye colour although their hooves and nose are less pigmented than usual, looking orange rather than black. Some are white from birth but others are born with a sandy coat but become progressively paler over several moults. All the varieties interbreed freely. Spots are genetically dominant to non-spotted, black pigment is dominant to brown pigment. A spotted deer with black pigment appears as a common colour but one with brown pigment appears as a menil. Black and brown are dominant to white. So, for example, two common-coloured parents could produce a fawn of any colour but two white parents could produce only a white fawn.

Another pelage variant is seen in Mortimer Forest, Shropshire, where some of the fallow are long-haired. These deer may be of any of the colour varieties but are distinguished by tufts of long hairs in the ears, curls on the forehead and longer body hairs.

Rear ends are especially useful when identification has to be made of a fast disappearing deer. The fallow deer's tail, which serves as a fly switch in summer, is longer, at about 28 cm, than that of any of Britain's other deer although only slightly longer than that of a sika. However, sika have a distinctive white rump patch which is flared like a giant powder-puff when the deer is alarmed. Markings shaped like an inverted horseshoe are conspicuous on the rump of menil and common-coloured fallow.

When going about their normal, undisturbed activities, fallow mostly walk or trot, but when faster progress is necessary they may run, bound or gallop. In addition they have a distinctive movement, a stiff-legged bounce on all four

Common, menil, white and black fallow bucks; note the distinctive black tail stripe and horseshoe pattern on the rump of common animals compared to the much lighter markings of the menil buck on the left.

feet, known as pronking or stotting. This is used when the deer is suddenly disturbed, although sometimes sheer *joie de vivre* appears to be the only explanation, especially when a doe and fawn pronk after a playful chase.

The sizes and weights of fallow deer show considerable variation, which reflects not just age and sex but also food availability and other habitat conditions. For example, the weights of adult females (does) range from around 35 kg to 55 kg. For males (bucks) older than 2 years, the range is mostly

between 53 kg to 90 kg, although some bucks over 110 kg have been recorded. These are whole body weights: stalkers usually refer to the weight of the body less internal organs or with head and feet also removed, so care is required when comparing information from different sources. A carcass, without the viscera, head, and feet, but with skin still on, usually weighs just under 60% of the whole body weight. The maximum body weight in wild or park populations is generally achieved by five years for a male and three years for a

Menil fallow such as this rutting buck retain their spotted coat markings throughout the winter.

6

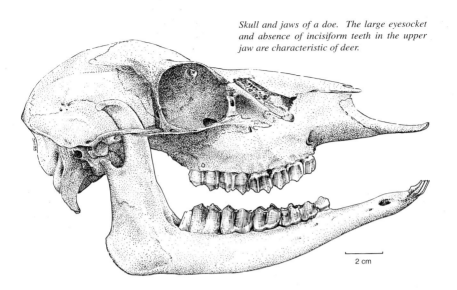

Skull and jaws of a doe. The large eyesocket and absence of incisiform teeth in the upper jaw are characteristic of deer.

2 cm

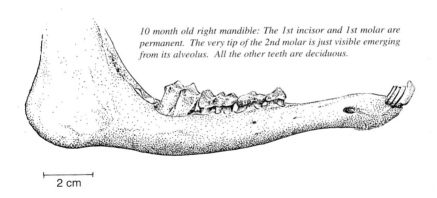

10 month old right mandible: The 1st incisor and 1st molar are permanent. The very tip of the 2nd molar is just visible emerging from its alveolus. All the other teeth are deciduous.

2 cm

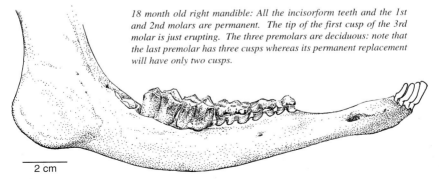

18 month old right mandible: All the incisorform teeth and the 1st and 2nd molars are permanent. The tip of the first cusp of the 3rd molar is just erupting. The three premolars are deciduous: note that the last premolar has three cusps whereas its permanent replacement will have only two cusps.

2 cm

female, but may be reached earlier with intensive feeding on farms. A typical height at the shoulder for a buck is 90 cm, with does about 10 cm shorter. The distance from nose to the last vertebra measured along the mid-dorsal line is around 160 cm for a buck, but nearly 20 cm less for a doe, though considerable variation occurs.

Antlers

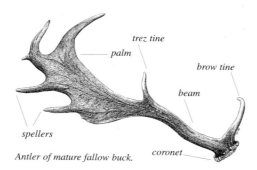

trez tine

palm

brow tine

beam

spellers

Antler of mature fallow buck. coronet

The adult males are distinguished by antlers: the characteristic shape of an antler on a full grown fallow buck is very different from any other deer in Britain because the distal end takes the form of a broad, flattened surface - known as the palm. The age at which palmation is achieved is variable. Fullest palmation may not be reached until 7 years of age or later, but as in any deer species, there is much individual variation in antler growth. Nutrition is of key importance in determining antler size achieved by a given age, but genetics affects the form of the antlers.

When a male fawn is about six or seven months old a pair of skin-covered bumps begin to grow from the frontal bones on either side of the skull. These are the pedicles which will be permanent structures. Some weeks later antlers begin to grow on the top of the pedicles. Antlers are initially formed in cartilage which gradually becomes mineralised to form bone. While growing they receive an internal blood supply via the pedicles. Externally the growing antlers are covered by a protective skin which also has a rich blood supply. This skin is known as velvet because of its texture created by densely packed short erect hairs. When the antlers have reached full size, the velvet is shed, the internal blood supply is cut off, and the deer is then said to be clean or in hard antler. The growing period usually lasts 12 to 17 weeks before the velvet is shed by the end of August. The shedding is a messy process during which, for some hours, pieces of bloody velvet dangle over the face until removed by rubbing vigorously against vegetation.

Antlers are cast and then re-grown each year, generally being replaced by a larger set until the buck reaches his prime. Most fallow bucks cast their antlers in April or May, although sometimes this occurs a month or two earlier. The timing of the events within the annual cycle is under hormonal control. The

Common fallow pricket among bluebells, showing single spike antlers typical of one-year old males.

development of the pedicles coincides with a surge in testosterone circulating in the blood but this level decreases while antler growth progresses. Another increase in testosterone causes hardening and cleaning of the grown antler and maintains it as dead bone attached to the pedicle until the next spring, when a drop in testosterone level causes the antlers to be cast. This cyclical pattern continues throughout the life of the buck.

Typically, the first pair of antlers are stubby knobs or simple spikes grown by yearling males, which are commonly referred to as prickets. Very occasionally wild males may carry one or two additional points and more complex first antlers occur on some deer in parks and farms. The second pair of antlers show the basic pattern of all future antlers. They consist of a main beam with a brow tine, arising just above the base of the antler and projecting forwards and upwards, and a second (trez) tine which rises higher up the beam. (Fallow deer lack a bez tine which in red deer grows at an intermediate point between the brow and trez tines). The distal end of a second antler bifurcates to varying degrees and may or may not show the beginning of palmation: in some individuals palmation is still lacking in the third head of antlers. Whilst some bucks achieve very broad palms (e.g. 18cm wide) in subsequent antlers, others have much narrower palms and some divide into a fish-tail formation. A variable number of short points, called spellers, protrude from the rear edge of the palm. For most bucks their most impressive antlers will be either the eighth or ninth pair.

Antlers are of importance for ritualised displays and sparring matches which enable bucks to assess each other and establish a hierarchy. Serious duels are begun only when both bucks are equally matched and each has a reasonable chance of winning.

Fallow bucks showing early re-growth and bifurcation of new antlers during early May.

The antlers of this buck are fairly typical of those of two to three year old males, showing no significant palmation.

For people who collect antler trophies, there are several schemes for scoring the standard of fallow antlers, the one internationally accepted in Europe being that of the Conseil International de la Chasse. Measurements are taken of the total length, the length and width of the palm, circumference of the coronet and at

Common fallow with very broad palmation and showing well developed spellers, brow tines and trez tines. Bucks with antlers of this size are rarely found in the wild in Britain due to heavy culling pressure and poaching.

two points along the beam. The weight of the antlers is also scored and additional beauty points are awarded for colour, the number of spellers and formation. Penalty points can be deducted e.g. if the span is less than 85 per cent of the average length of the beam. A score of 180 or over rates as a Gold Medal head, 170 for Silver and 160 for Bronze. Such specimens are displayed at international Game Hunting Exhibitions where the major winners are usually from east European countries within which the opportunity of shooting deer with massive antlers is an important source of revenue.

Use of scent

Like most deer, fallow are well endowed with scent glands. Undoubtedly their secretions are important in conveying information to conspecifics although the human understanding of their meaning remains limited. On the hind legs, just below the hock, a raised pad of hairs can be seen, although this is not as conspicuous as in sika deer. This is the site of the metatarsal gland that produces a cream-coloured waxy secretion which smells rather like rancid butter. This scent is individually distinct, but does not appear to contain specific information about age, sex or rank in fallow deer.

Below the corner of each eye is a small fold in the skin which houses the sub-orbital gland, which fits into a small pit in the skull immediately in front of the eye orbit and produces a brown waxy substance. This gland appears to be particularly active in fallow males during the rut, when a white milky secretion may be discharged, staining the cheeks.

Between the two cleaves of each hind foot is a deep pocket in the skin. This is lined with the interdigital glands which secrete a substance used for marking the ground. These glands too appear to be most active in bucks during the breeding season, when scrapes are made around fixed mating territories or 'rutting stands' and on paths regularly used by does, to intercept them as they move to and from feeding grounds.

In addition to these glands, males have one further exceptionally smelly, seasonally active part of the body. At all times the tassel of long coarse hairs (brush) which hang from the penis sheath is obvious and is useful in distinguishing at a distance between young males and females. However, for the duration of the rutting season the tip of the penis sheath everts and the surface looks like cracked, baked soil. The tassel hairs splay outwards and become heavily stained, as does the groin. Very large sebaceous glands on the everted prepuce seem to be the source of the characteristic rutty buck smell, pungent even to a human nose.

Vision

Grazing animals, ever watchful for predators, need a wide angle of view. Eyes positioned on the sides of the head, horizontal pupils and a downward sloping face all contribute to this. Several features in the structure of the eye of a deer equip them with vision superior to that of humans in low light conditions. The retina of the deer has more rods (cells for dim light vision) than cones (cells which enable good daytime and colour vision). The pupil can be opened very wide and at the back of the eye is a mirror-like layer (tapetum) which reflects back to the retina any light that was not absorbed the first time. This layer causes the green eye shine at night when deer are viewed in a beam of light.

Fallow deer are thought to see a more blurred image than humans and within a restricted colour range. The former has long been surmised from the habit of fallow deer to peer, extending the neck and moving the head from side to side thus trying to get several views of a strange object or intruder before deciding whether it needs to flee to safety. Recent electrophysiological measurements in the retinas of white-tailed and fallow deer indicate that they are dichromatic, having only two types of cones (rather than three as in humans), the type sensitive to long-wave colours (red, orange) being absent. When it comes to clothing for deer watching or stalking, the main criteria are that the garments permit silent movement, are not shiny and therefore reflective, and preferably have a broken pattern which is less noticeable than an expanse of one light colour. Nevertheless deer are adept at detecting movement, hence the need for the would-be deer watcher to make only slow, gradual movements, for example, when raising binoculars.

Vocalisations

Bucks seldom vocalise during most of the year but their groaning during the rutting season has to be heard to be believed: the gutteral belching groan may be heard up to a kilometre away and is a sound that is rarely mistaken or forgotten! At the peak of the rut a mature buck may emit groans as often as every two or three seconds for several minutes at a time. Whilst groaning the head is tilted back and the Adam's apple (larynx) can be seen to drop by about 15 cm.

During the rut females being chivvied often mew or whicker softly. An alarmed doe may bark at any time but this is most likely when she is accompanied by a fawn. Communication between doe and young fawn also includes pheeps and bleats, and tends to be used increasingly in addition to visual communication where fallow aggregate in large herds. A recent study in deer parks found that among observed interactions between mothers and fawns, 95% were primarily visual, 22.0% used vocal/auditory interactions (in addition or instead of visual signals) and 17.8% also employed scent signals. Adults looked for visual cues more frequently than responding to vocal or olfactory cues from the fawns; conversely, fawns listened and responded to auditory cues more frequently than adults.

The sound-gathering powers of the fallow deer are good. Their ears, separately or in unison, can be swivelled through about 180 degrees while keeping the head and body still and so avoiding drawing attention to themselves.

Doe with her well-grown fawn during early autumn.

Signs

Where fallow are resident or have passed through there will be a variety of signs. Usually the first clue will be the footprints (slots). Their clarity varies according to the type of soil and whether it is wet or dry. Usually just the two cleaves of a hoof make an impression but if the deer lands in soft mud the dew claws (vestigial toes above and behind the hoof) also imprint. As the deer walks the hind foot registers in the slot made by the fore foot. At a walking gait the slots will be approximately 35 to 50 cm apart depending upon the size of the deer. A good clear slot of a doe would be about 5 cm long, and that of a buck up to 8 cm. However, the prints of an adult roe and a young fallow, a sika and a fallow or fallow and young red deer could be confused.

Fallow are creatures of habit which use the same routes regularly, so well-worn paths, perhaps crossing the perimeter of a wood or across a woodland ride, may suggest suitable vantage points from which to observe the deer. Although capable of jumping a two-metre fence, fallow always prefer to push through or under a barrier. As they do so little bundles of hairs are often left on barbed wire or hedges.

The vigorous activities in which the bucks indulge in the autumn leave their marks for months, long after the bucks have moved out of the rutting area. Thrashed bushes are less conspicuous once their leaves have been shed but scrapes in the ground are visible until covered by fallen leaves or the growth of new plants. Frayed tree bark, mostly about 120 to 160 cm above ground level, is a relatively permanent indication that rutting activities have taken place.

Habitat

Fallow are primarily considered as inhabitants of mature broadleaf woodlands offering an abundance of grassy glades and well developed shrub vegetation. Whereas these elements combine to make up what might be thought of as their ideal range, the adaptability of fallow is such that populations thrive almost everywhere except in high mountains. The only region where fallow have persisted in their natural state is reputed to be the coastal wooded plains of southern Turkey.

Today populations of fallow occur in most woodland types throughout Britain, central and southern Europe, as well as in the many other parts of the world where they have been introduced. While deciduous broadleaf or mixed woods are preferred, fallow will also colonise conifer plantations provided these contain or border onto more open habitats. In Britain they are also increasingly found living in relatively open agricultural landscapes offering only small patches of tree cover, as well as in the traditional parkland settings from where most of our feral herds derive.

Woods inhabited by fallow need not be very large as they are used primarily for shelter, with the deer not normally dependent on the woodland's food supplies. As 'preferential grazers', fallow will tend to graze mostly on rides or ground vegetation between the trees, and forage on agricultural crops, pastures and other habitats beyond the woodland edge. Habitat use changes through the seasons as the availability and nutritional value of different forages alters, so that in agricultural areas the local cropping pattern may exert a major influence

in determining the seasonal ranging behaviour of the deer. Open fields, pastures and grass leys are used most during late winter and spring, when little forage is available within the woodland boundary. The cover of woodland, and of hay meadows and cereal fields, is sought increasingly from early summer when the young are born and use of woodland for foraging peaks during autumn when tree mast and other fruit become abundant.

Tree cover is also sought more during the daytime, with the deer likely to venture further from the woodland edge during the cover of darkness. However, by comparison with sika and red deer, fallow are less prone to shifting to nocturnal activity in response to human

Although able to utilise a wide range of habitats, woodland forms a very important part of the home range of most wild fallow deer.

disturbances. They can habituate remarkably readily to traffic noise or other regular activity, so that in many parts of England it is not unusual to see large herds of wild fallow grazing or resting during the middle of the day, even in fields directly abutting major motorways.

Food and feeding behaviour

Fallow are mixed or intermediate feeders which by preference will feed predominantly on grasses for most of the year. However, the structure of their rumen, relatively large row of incisors but fairly narrow muzzle combine to enable fallow to feed more selectively when required, and thus to utilise a wide variety of other forages when availability and digestibility of grasses falls during autumn and winter. Most dietary studies have indicated that grasses comprise between 60% and 95% of fallow diets from spring to autumn. Other

Wild common doe in winter on cereal field; note the long guard hairs which extend well above the main coat.

foods eaten range from acorns, beech mast, chestnuts, crab apples, fungi and berries taken especially during the autumn and early winter, to a wide variety of different herbs, cereals, legumes and browse (leaves and twigs) including bramble, holly, ivy, heather, conifers and tree bark. The diet of male and female fallow may differ to some degree. In the New Forest in Hampshire, for example, does have been found to take a higher proportion of grass but less rushes, sedges, dwarf shrubs and browse than bucks.

Feeding activity usually alternates with periods of rumination at around three to four hourly intervals throughout the day and night, but the most intensive and longest grazing bouts tend to occur during early morning and around dusk. In summer, when nutritional demands for lactation and skeletal growth are at a peak, as much as 80% of the day may be spent in feeding activities. Fewer feeding bouts occur during winter. Food

Wild buck in winter coat on stubble field.

intake at that time of year is reduced through what is termed winter inappetance: such reduction in food intake, with lack of any significant growth or weight gain over the winter, is usually exhibited even if food is made available *ad libitum*.

Under the persistently high stocking conditions normally maintained in deer parks, all shrub and understorey vegetation and tree branches within reach of deer quickly become depleted, creating a browse line. Supplementary feeding therefore becomes essential in most parks over the winter months when digestibility and nutritional value of the grass sward falls.

Competition with other deer

Wild fallow do not hybridise with any other deer in Britain, but often share their habitat with one or more of our other species, most commonly muntjac and/or roe deer. Neither of these competes for food to a significant degree with fallow, because they are 'concentrate selectors', selecting mainly just the most easily digested and nutritious plants or plant parts, and therefore are more dependent than fallow on the availability of diverse shrub and herb layers. High densities of fallow can sometimes deplete woodlands of any significant re-growth of such understorey and therefore may make certain areas less suitable for smaller deer species. This competition is unidirectional, in that fallow themselves will rarely be affected to a significant extent by the browsing activities of the smaller deer species.

Fallow will also live readily alongside red deer and sika, with which there is more dietary overlap. In several areas of England, such as Thetford Forest in Norfolk and the New Forest in Hampshire, fallow deer thrive alongside significant resident populations of at least three and sometimes four other deer species. Fallow are also regularly kept together with red deer, and sometimes with sika and other exotic species, in deer parks. In general fallow seem to perform rather better in terms of their body weights and reproductive success in those parks where they are alone, rather than when together with other deer. In terms of their grazing style fallow should be able to out-compete red deer in such mixed parks, as their smaller muzzles allow closer cropping and hence better utilisation of heavily grazed swards. The reasons for their lesser performance in mixed species parks is hence more likely to relate to some degree of aggressive displacement by, for example, red deer from access to supplementary feeding and/or favoured grazing areas.

Social behaviour

Fallow are often described as being a herding species, and indeed at favoured feeding or resting places in the open are sometimes found in large aggregations which may number from 30 - 200 individuals. Such large gatherings are however fairly transient, with differing individuals regularly leaving or joining the herd, and commonly splitting up into much smaller group sizes on moving into woodland or other habitats. The largest stable groups commonly consist of merely five or fewer individuals, made up of one or two adult does with their current and sometimes a previous year's offspring.

In large fallow populations, adult bucks live segregated from females and young for much of the year, and form fairly fluid 'bachelor groups' of unrelated individuals for varying periods. The degree of sexual and spatial segregation is very variable in differing environments. In many populations bucks remain in female areas only during the autumn breeding period, and then move to distinct geographical ranges. Thus, in studies made within the New Forest in Hampshire over 95% of groups encountered between December-September were either exclusively male or else does, fawns and prickets (bucks less than 24 months old). However, in some populations adult bucks remain with doe herds for longer after the rut, sometimes as late as April-May, whereas in others again, particularly those living in more open landscapes, aggregations containing adults of both sexes are comparatively common throughout the year.

Adult bucks gradually become more aggressive to one another once antler cleaning is completed during August, which soon leads to the break-up of any bachelor groups. Bucks then move into doe areas during early autumn and compete against other males in attempts to set up display grounds and call to attract does. Groups containing fully grown adults of both sexes are observed most frequently during October to December. The rutting groups then break-up and adult males drift off to re-establish bachelor groups or may remain solitary.

The size of group in which fallow will most commonly be encountered is also influenced by habitat type and season. In general, sizes of doe groups (including young of either sex up to 20 months of age) within large woodlands have been found to average merely four during winter and spring and two during summer. However, in more open habitats average group sizes are often between two and four times larger. Males tend to be more solitary than females, with groups of more than six adult bucks seen only quite rarely in the wild throughout the year. Nevertheless, in some of the large fallow populations, such as found in the Wyre Forest, Cannock Chase and the New Forest, all-male aggregations of 25 to over 100 have sometimes been observed.

When observing a female or mixed-sex herd a hierarchy led by a dominant doe will sometimes be apparent. On being disturbed the dominant doe will usually lead the group away, followed by younger does and fawns, with any bucks usually last to leave. Social interactions outside the rut, such as mutual grooming, are relatively rare, except between does and their fawns.

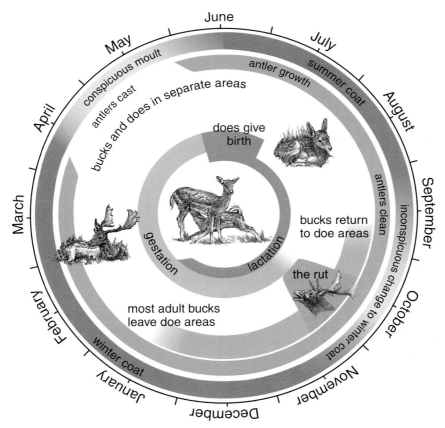

The annual cycles in fallow deer.

Mating behaviour

Fallow deer are seasonal breeders. In Britain the peak period for fallow mating activity, the 'rut', usually occurs between mid-October and early November. The fallow rut can provide a particularly fascinating spectacle including a wide array of ritualised displays and fighting behaviour. The mature bucks also develop a number of secondary sexual characteristics during this time. Most obvious of these are a large increase in the girth of the neck, eversion and staining of the end of the penis sheath and staining of the groins accompanied by a pungent, lingering, rutting odour. The normally quiet and placid bucks also become increasingly aggressive towards one another, and develop a deep loud belching groan, which once heard is unlikely to be to be mistaken for any other species.

White fallow, showing the prominent staining of the brush and flanks noted among mature bucks during the rut.

In most populations the adult males move into the female areas from late September onwards. Here they may compete with other males for rutting stands. Bucks display their presence on a rutting stand both to females and to competing males through frequent loud groaning. They will also create scrapes in the ground by pawing them with their feet, and mark the hollow by urinating. Bushes and nearby tree branches may also be thrashed and marked by rubbing the sub-orbital scent gland against them. Contests with other males will often be resolved without physical contact merely through posturing and strutting along the boundaries of the rutting stand, but direct fights are not uncommon. Before a fight ensues other ritual displays may include thrashing of vegetation, and 'parallel walking'; during the latter the two bucks walk about 5 to 10 metres apart eyeing their opponent with heads tilted slightly to one side ready to react to any move. Eventually the animals may turn suddenly, clash and lock antlers crown to crown with the nose pointing down to the ground, and attempt for several minutes to shove their opponent backwards in a trial of strength. Serious fights usually occur only between animals of comparable size, but deaths and serious injuries may occur especially towards the later stages of the rut, when those bucks most active at the peak of the rut start to lose condition and are displaced from their stands by fresh males.

The setting up of discrete widely separated rutting stands had long been thought to represent the typical behaviour pattern for this species. However, detailed studies comparing a wide range of populations over the last two decades have shown that fallow exhibit extreme variation in their mating system. In common with other polygynous species, the great majority of matings each year will be attained by just a small proportion of the mature bucks. Often as many as three quarters of fawns are fathered by less than 10% of the adult bucks present in any year, irrespective of the mating system. However, in order to maximise their own rank in this highly skewed distribution, individual bucks may adopt a wide range of different strategies given differing population structure and environmental conditions.

While bucks in some areas or years establish discrete rutting stands, as described above, others may defend only very temporary stands and thereafter consort with the harem they have attracted. In other populations, particularly where only a few bucks of similar size are present, bucks may compete for neither territories or harems: males and females instead mix in large mobile mixed sex herds and the largest bucks attempt to gain overall dominance within the herd, with dominance rank conferring right of access to oestrous does.

Two bucks fighting during the rut.

Where rutting territories are established, these may be either widely separated from those of other bucks, or may be contiguous in clusters of several adjacent stands. In the extreme, as many as five to 25 contiguous territories have been observed to be defended, some merely 5 to 10 metres across and entirely surrounded by territories of other bucks. This situation is directly analogous to the leks formed by birds such as black grouse. Fallow deer are one of the few ungulate species where lek breeding has been clearly demonstrated. However, even in those populations where they have been observed leks are not necessarily formed every year. The type of mating system exhibited is influenced especially by the density of mature males, degree of aggregation and density of females, and is associated also with the structure of the habitat.

Buck creating a 'scrape' of bare ground on his rutting stand.

Locations used for rutting may be very traditional and the same bucks often return to the same area in successive years. Does may also be faithful to a particular rutting territory year after year and are often accompanied by their daughters. Although this makes it feasible, in theory, for a buck to mate with his own daughters, the probability of this occurring among free ranging populations is fairly low. In order to mate with his own daughters a buck would need to maintain a high reproductive rank in three or more consecutive years. In reality, in the majority of populations, turnover of the most successful breeding males tends to be much more rapid than this.

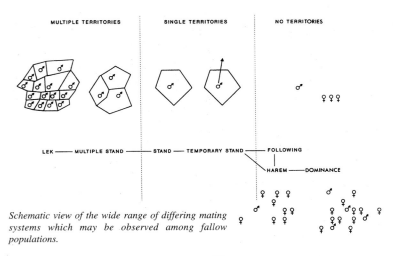

Schematic view of the wide range of differing mating systems which may be observed among fallow populations.

Once several does have settled on a rutting territory, the buck frequently herds them to maintain a cohesive group, and periodically moves among them nuzzling either their neck or anogenital region. Immediately before copulation, the male will follow the doe more persistently at a slow walking pace, often tilting his head to one side, groaning, nuzzling and pushing his muzzle against her flanks. Copulation may be preceded by three to 20 or more attempted mountings. Does remain receptive for only a brief period and only a very low proportion of does are mated by more than one buck during the same oestrous cycle.

During the rut bucks will feed very little, though they occasionally will leave their stands to find water, and may lose more than a quarter of their initial autumn body weight by mid November.

Menil buck mating on rutting stand; the master buck may tolerate some younger males such as here on the right provided they don't approach too closely.

Rearing fawns

Gestation lasts about 229 days. Fawns are born during June or early July, but may at times be born as late as November to females that conceive after the rut. During early summer good foraging is available to mothers to sustain the extra demands of lactation. Cover such as nettles, bracken and grass also reaches heights sufficient to conceal and provide shelter for newly born fawns at this time, which is important because fawns are left unattended for much of their first few weeks of life. The mean birth weight of healthy fawns is 4.5 kg. However, those born to first-time mothers conceiving as yearlings tend to be lighter, averaging 3.7 kg, and are conceived and born an average of 11 days later than those born to older dams.

One week old common coloured fawn.

A single fawn is the norm. Although twin and even triplet foetuses do occur, the live birth of more than a singleton is very rare. Newly born fawns rely on their cryptic coloration for camouflage as much of their first three weeks is spent lying in bracken, leaf litter, long grass or shrubs. During this hiding phase, the mother returns to suckle the fawn around five or six times per day for the first three weeks. The fawn gradually spends less time resting and more time following its mother which will have rejoined her social group. Young fawns may be seen at times in small nursery groups staying slightly apart or on the periphery of the main herd while the adult animals graze, with just one or two does staying near them. Young fawns indulge in play behaviour, such as sudden chases, jumps and pronking, particularly in the early morning and evening hours during the summer, and occasionally young does may also join in with the gambolling. The fawns grow rapidly, with vegetation increasingly supplementing milk and in time herbage takes over as the main food. However, if environmental conditions are good, many does continue to lactate well into, and sometimes throughout, the winter months. The dams may sometimes allow another dam's fawn to suckle, so two fawns seen with one doe is not proof of twins. Such allosuckling may be tolerated by the dam for a significant number of times in some small park or farm herds maintained at very high density;

During the first few weeks of life fawns bed down in cover for much of the day.

elsewhere the dam is usually observed to terminate the suckling bout quickly once she has detected the 'thief' as not being her own fawn.

Although fawns can be, and on farms commonly are, weaned by November, they do still benefit much from the continued protection of their dams over winter. The close contact between dam and fawn also teaches the fawn about the best areas for foraging and safe places for shelter, and a fawn that loses its dam during this learning phase is likely to be disadvantaged.

Home range

Outside of the rut, both bucks and does are non-territorial, but tend to stay within fairly small overlapping home range areas. Any one individual may cover an area from less than 1 to around 2 km² (100 to 200 ha) for much of the year. However, by comparison with red and roe deer, few detailed radio-tracking data are available on the ranging behaviour of fallow in different habitat types. Long-term direct observations of individuals identified by their spot patterns in the New Forest suggest summer range sizes of between 50 - 90 ha for females and a mean of 110 ha for adult males, with the ranges increasing by around 50% during winter. More recent studies in the East of England, using GPS satellite tracking techniques, have confirmed that individual range sizes are somewhat larger for fallow living in agricultural landscapes interspersed with only small woodlands, though mean year-round ranges still averaged merely 175 ha for females and around 200 ha for males. Personal observations suggest that future work may reveal that considerably larger ranges are utilised in some cases, as whole herds of fallow have been noted to move between widely separated copses in different seasons in some areas when disturbed by game-bird shoots or harvesting of agricultural crops on which the deer had been feeding.

Population densities and dispersal

The fairly undemanding dietary habits of fallow compared with species such roe or red deer, enable populations to build up to relatively high densities before significant levels of damage to the habitat and planted trees become apparent. Their own social tolerance of sometimes living in very large groups, and consequently very low dispersal rates, often results in the build up of extremely high numbers at a local level. While most feral fallow populations will occur at around 5-20 deer/km², densities of 30-75 per km² or even higher are not uncommon in some parts of England and Wales. This can lead to the woods being heavily overgrazed and depleted almost entirely of any understorey.

In traditional deer parks, fallow herds will often be maintained year-round at well over ten times the normal density at which they are found in the wild state, with most parks stocked to at least 50 and sometimes up to 800 deer per km^2. With more intensive pasture management, year-round supplementary feeding and housing over winter, deer farms sometimes raise stocking levels still further to around 10-20 per hectare (1000-2000 per km^2).

Information on the dispersal behaviour of fallow remains sparse, but their spread and natural colonisation of new areas is usually a very gradual affair. Most dispersal movements are of juvenile males which may leave their family groups in spring just prior to birth of the next fawn. The dangers of dispersal are emphasised by the fact that although sex ratios in most managed populations are strongly female biased, a much higher relative proportion of males is recorded as casualties of road traffic accidents. Fallow deer are most often seen crossing major roads from February to April, coinciding with dispersal of males, and between September to November associated mainly with the increased movement of bucks to and from rutting areas.

Population condition and dynamics

Although able to adapt to living across a wide range of latitudes and habitat types, the mean body size and reproductive rates of fallow show much variation between populations. Large differences in body weight achieved by fallow have been noted between differing park and wild populations in England and Wales, which can be largely attributed to differences in population density, habitat conditions and climate, rather than genetics. For example, in a sample of 15 deer parks in England and Wales, mean winter live weights of male fawns were as much as 58% higher in some parks than in those with the lightest deer, with differences up to 30% also noted between average weights achieved by adults. Such differences between populations often relate foremost to the quality and productivity of pastures over the summer, and feeding regime over winter. However, even within single populations, body weights may vary widely in direct relation to changes in stocking levels, with the population density (or resources available per animal) during the year of birth being even more influential on final adult weights than density in later years.

The breeding success of females is also closely related to environmental conditions. At high densities where food availability is low, the proportion of does conceiving can fall sharply. The proportion of does conceiving as yearlings (i.e. around 16 months old) serves as a particularly sensitive indicator of population condition, and may range from over 90% in some to less than 20% in other park populations. Pregnancy rate among older does is less variable with rates commonly between 75% and 95%. The likelihood of conceiving during the rut is closely related to body weight: few yearling does become pregnant unless they have reached a live weight of 32kg by the end of autumn.

Mortality rates tend to be highest during the first few months of life, but in general survival rates of fallow fawns are good in comparison with roe or sika. Although losses are probably less than 10% during most summers, mortality of fawns may escalate sharply during prolonged wet weather, especially in locations lacking dense cover for shelter.

Predation and mortality

Britain lacks large carnivores, so the only natural predation on fallow deer is by foxes which are capable of taking very young fawns. Dogs allowed to run loose may also maim or kill young fallow.

The longevity record for a wild fallow deer seems to be 16.5 years, although a few years more have been reported in captivity. Most populations are culled to some extent so very few individuals will reach such an advanced age. In a study of 87 fallow which were found dead or debilitated in Essex (mostly as result of road traffic accidents), 37% had died by two years of age whereas only 16% were seven years or older.

Aside from deliberate culling, road accidents can take a very large toll in some areas. Close to 100 such accidents are recorded annually near each of several major fallow deer forests in England. Firm nation-wide statistics of numbers of deer killed in road traffic accidents remain lacking for the UK, but are likely to lie well in excess of 20,000 and possibly over 40,000 per annum for all our species combined. Of these several thousand collisions will involve fallow. Accidents with fallow bring the added welfare problem that they, like red and sika, are more than twice as likely to require humane dispatch at the roadside as compared to muntjac and roe, which are more likely to be killed outright in collisions with vehicles. Too many drivers fail to heed roadside warning signs, not realising the potentially very serious results of an impact. High tensile roadside fencing remains the only well-proven means of reducing deer accidents, especially where combined with provisions for alternative passage of deer via underpasses or overbridges of appropriate dimensions. The evidence for the efficacy of reflectors, designed to reflect car headlights to forewarn the deer of oncoming traffic, at reducing deer accidents remains equivocal, and can at best only deter deer from crossing roads during hours of darkness.

Other miscellaneous causes of death include being hit by trains, caught on fences (fallow are particularly prone to tripping between the two top horizontal line wires on stock fences, leaving the animal hanging by one leg) or starvation because two bucks have inextricably interlocked antlers during a fight. Limb bones and vertebrae do often show signs of traumatic injuries, with irregular growth of new bone round fracture sites.

Parasites and disease

Ticks, lice and keds are the three ectoparasites most commonly encountered on fallow deer although the incidence varies in different populations. Adult female ticks which have engorged on the blood of the deer are most easily spotted on the least hairy parts of the skin when a carcass is examined. They look very much like a castor oil seed or a small brown grape. Less easy to spot are the tiny larvae and the nymphs and males. Although called the deer or sheep tick, this species will feed on a wide range of hosts, including humans. A high proportion of ticks are infected with the organism which can cause Lyme disease in humans. General malaise, or influenza-like symptoms sometimes accompanied by a red ring spreading out from the site of a tick bite, are among the indications that medical advice should be sought. Antibiotics usually enable full recovery if spotted early, but a delay in treatment may have serious consequences in the future, perhaps even years later.

Lice are so tiny that they will seldom be noticed unless the deer is ailing and has an unusually heavy burden. If viewed with a hand lens they will be seen clinging to hairs or creeping over the skin particularly in the groin. Whereas lice are wingless insects, deer keds have wings but these break off soon after the ked has landed on its host. These squashed-looking flies scuttle about, crab-like, among the hairs, feeding on blood from the skin.

The incidence of liverfluke and lung-worm varies in different localities but is probably rarely a problem to the deer. Tape worm cysts may be present, looking like a small, fluid filled ping-pong ball attached at a number of possible sites within the abdominal cavity, but can not develop to the worm stage within the deer. Although fallow deer are susceptible to various infectious diseases such as yersiniosis, TB (avian and bovine) and Foot and Mouth Disease, the incidence of disease in wild populations in Britain seems to remain very low.

Impact and management

Of all our deer, fallow have been most closely associated with people in Britain ever since they were re-introduced to Britain by the Normans. Initially fallow were kept primarily for hunting but also to adorn the parks surrounding the homes of noblemen. Numerous 'forests' such as the New Forest and Epping Forest, which continue to support important fallow populations today, were first established as royal hunting reserves for the Norman Kings. Such areas, together with the numerous fenced deer parks established throughout England and to a lesser extent Scotland and Wales, have led to the very wide distribution the species enjoys in Britain today.

In deer parks, fallow may become accustomed to people ... provided dogs are kept under control.

Besides their amenity and sporting value, the presence of high numbers of fallow can have negative impacts especially for forestry and agriculture, and in some cases on conservation habitats. In woodland fallow may damage young plantings or prevent regeneration of coppice stands through browsing. Bark stripped from mature deciduous and coniferous trees can also be a problem, particularly in hard winters. Although fallow are less prone than roe deer to cause browsing damage to young trees, their greater height allows adults to browse to 1.4 m, and some may reach higher by rearing up on their hind legs. Tree growth shelters need to be at least 1.6 metres high to be effective against fallow, but even then fallow sometimes develop a habit of thrashing and knocking over such shelters with their antlers and/or feet. Significant damage by fallow to agricultural crops is less widespread, but can become highly significant locally. Indeed, in several parts of East Anglia, 70 to 200 fallow regularly congregate on a few favoured fields. Such high densities can also pose problems for some conservation areas. Recent declines in oxslips in some ancient woodlands in East Anglia, for example, as well as an inability to re-establish coppice woodlands, have been blamed, rightly or wrongly, on the coincident proliferation of fallow deer throughout that region.

To maintain fallow numbers at sustainable levels the main method of control is shooting with a rifle. Walked shotgun drives, as practised in the past, were made illegal in 1963. Hunting of fallow on horseback while following a pack of hounds was last practised, in the New Forest, in 1997. By culling at an appropriate level, which commonly lies somewhere between 20 to 35 % of the numbers present in autumn, fallow populations can be kept in check whilst also producing an annual harvest of venison and revenue from stalking.

Buck rearing up on hind legs to reach foliage. Note the well developed 'browse line' caused by deer at high density.

A 'bachelor' group of bucks during summer.

Culling alone is rarely sufficient to reduce or prevent damage. In practice, effective control of numbers and impact will generally only be achieved via integrated management, involving culling to control numbers combined with non-lethal means to reduce the impact of the remaining deer. Impact may be reduced through use of permanent or temporary fencing, tree guards and shelters, chemical repellents and habitat manipulation. Culling may be further facilitated by placing high seats along a system of wide rides within or on the edge of woodland. In central Europe, where winters are harsher than in this country, wild deer populations are provided with supplementary feed during winter, to help in alleviating browsing activity on planted trees and agricultural crops. In this country, especially in the case of fallow, raising the natural forage supply within the woodland boundary through developing 'deer lawns' or grassy rides can be similarly effective by drawing the deer away from vulnerable trees and crops.

Fallow in snow showing duller coat with fainter spot patterns.

Grass swards alone cannot usually sustain fallow over winter in this country. Growth of grass slows during early autumn and usually ceases altogether when daily mean temperature falls below 5°C, and digestibility of the remaining sward is also much reduced during winter. In most deer parks, where little alternative forage aside from pasture is usually available, supplementary feeding over the winter months is therefore essential. As a precaution against high late winter mortalities it is advisable to start feeding from mid-November until early March, ideally at a level sufficient to satisfy around 80% of their basic daily metabolic energy requirements (i.e. to provide approximately 9 megajoules of metabolisable energy per day). To delay feeding until severe weather arrives risks depletion of most body fat reserves early on in winter, with consequent rapid loss of condition if a cold spell does occur.

Farming

Deer farming in Britain took off seriously less than 30 years ago, focusing mainly on red deer in Scotland. Farming of fallow began even more recently during the 1980s, and by 1998 fallow made up 20% of approximately 25,000 deer farmed in England and Wales, with very few fallow being farmed in Scotland. By contrast, on the continent, fallow are farmed in much greater numbers than red deer. Fallow is highly regarded as the best of venison and the size of the joints suits many domestic customers better than the larger cuts from red deer. Deer meat in general has gained increasing favour because it is low in cholesterol, is normally free of additives and the husbandry is regarded as welfare-friendly.

Fawns on farms remain in the paddocks with their mothers until they are weaned in the autumn and are then commonly over-wintered indoors before

Does in good condition may continue to suckle their fawns occasionally for much of the winter.

being released on to fresh pasture in the spring. Apart from individuals selected as breeding stock, most fallow deer on farms are killed at between 15 and 24 months of age, taking advantage of the period of rapid growth and ensuring that all farmed venison is tender.

Legal status

Anyone involved with managing fallow deer should be conversant with the relevant legislation, especially The Deer Act (1991), which applies in England and Wales, the Deer (Scotland) Act (1996), and the Wildlife and Countryside Act (1981). These Acts make it illegal to kill a deer using a rifle with a calibre of less than 0.240 inches or a muzzle energy of less than 1700 ft. lbs. They also prohibit the use of any airgun, cross bow, net, trap, or stupefying drug for taking deer (unless licensed under special exemption for scientific purposes).

Mixed sex herd on agricultural land during the winter.

Deer may also not be culled during night-time or in the close seasons laid down in the above Acts, which differ between species as well as between Scotland and the remainder of the UK. At present the close season for fallow males lasts from 1 May to 31 July, and for females from 1 March to 31 October (16 February to 20 October in case of Scotland).

Further reading

Comprehensive accounts of all British mammals can be found in *The Handbook of British Mammals* 3rd edition edited by G.B. Corbet and S. Harris. Blackwell Science, Oxford, 1991. (Out of print)

More detailed information on fallow deer is provided in the following books:

Fallow Deer: their History, Distribution and Biology by D. and N. Chapman. Coch-y-bonddu Books, Machynlleth, 1997.

The Deer Manager's Companion: a guide to the managment of deer in the wild and in parks by R.J. Putman and J. Langbein. Swan Hill Press, Shrewsbury, 2003.

The Natural History of Deer by R. J. Putman. Christopher Helm, London, 1988.

Further background on other deer species can be found in *Deer of Britain and Ireland: their Origins and Distribution* by P. Carne. Swan Hill Press, Shrewsbury, 2000.

Useful addresses

The Mammal Society, 2B Inworth Street, London SW11 3EP.
Tel: 020 7350 2200 Fax: 020 7350 2211
Web site: www.mammal.org.uk; email: enquiries@mammal.org.uk

The British Deer Society, Burgate Manor, Fordingbridge, Hampshire SP6 1EF
Tel. 01425 655434, Fax. 01425 655433
Web site: www.bds.org.uk; email: h.q@bds.org.uk

The Deer Initiative (for co-ordination and advice regarding deer management groups, etc.) PO Box 260, Bridgwater TA5 1YG
Web site: www.thedeerinitiative.co.uk; email: info@thedeerinitiative.co.uk

The Deer Commission for Scotland, Knowsley, 82 Fairfield Road, Inverness IV3 5LH Web site: www.dcs.gov.uk; email: deercom@aol.com

Acknowledgements

Over many years the authors have appreciated the co-operation of many individuals, authorities and organisations in the course of their studies of fallow deer. Rory Putman and Alastair Ward also kindly commented on a draft of this booklet, Sarah Wroot drew the illustrations, and a number of others (individually shown on Page 2) provided additional photographs for this booklet. Sincere thanks are extended to them all.

British Deer Series

Produced by The Mammal Society in association with the British Deer Society

Other books in the series include:

Muntjac £3.50

By Norma Chapman and Stephen Harris

Reeves' muntjac is a small, primitive, deer native to China and Taiwan but is widely established in the wild in Britain. Norma Chapman and Stephen Harris have studied them for over 20 years and provide an interesting guide to their biology and ecology.

Roe Deer £3.50

By John Fawcett

Roe deer are one of only two species of deer native to Britain. They are also the most widespread deer found here. John Fawcett has been studying them for many years and reviews their biology and ecology as well as advising the best way to observe them in their natural environment.

Chinese Water Deer £3.50

By Arnold Cooke and Lynne Farrell

Water deer lack antlers but the upper canine teeth of the bucks are impressive A few populations of this deer are established in England following escapes or releases from captivity. Arnold Cooke and Lynne Farrell have been studying them for over 25 years and discuss their biology and ecology in this country.

Sika Deer £3.50

By Rory Putman

One of three Asiatic species of deer found in Britain, sika deer were originally introduced to Ireland in 1860. Following escapes from several parks, sika are now established in a few areas of England and Ireland and are increasingly widespread in Scotland. They are closely related to red deer with which they can interbreed. Rory Putman introduces this fascinating species and the impact they have had in Britain.

To order any of these books contact The Mammal Society on 020 7350 2200 or The British Deer Society on 01425 655434.